BLUEPRINT OF A SYNTHETIC ORGANISM:

The Fusion of Structure, Hybrid Energy, and Decentralized Intelligence

Blueprint Of A Synthetic Organism: The Fusion Of Structure, Hybrid Energy, And Decentralized Intelligence

by
Sulien Valentino Solovyov

ISBN: 978-1-989647-90-5
First published November 8 2025
Toronto, Canada

Publisher: The Evergreen Centre

Title: Blueprint Of A Synthetic Organism: The Fusion Of Structure, Hybrid Energy, And Decentralized IntelligenceSolovyov, Sulien Valentino.   Blueprint Of A Synthetic Organism: The Fusion Of Structure, Hybrid Energy, And Decentralized Intelligence   — First edition.

Summary: A conceptual blueprint and technical roadmap for next-generation machines, detailing a unified architecture that merges structural load-bearing materials, continuous ambient energy harvesting, and fully decentralized computational intelligence to create fault-tolerant, self-sustaining robotic and infrastructural systems.   Identifiers: ISBN 978-1-989647-90-5 Classification Subjects (Library of Congress)1. Robotics—Technological innovations.2. Bio-inspired engineering.3. Hybrid energy harvesting.4. Decentralized control systems.5. Smart materials—Design and construction. Dewey Classification 629.892—dc23

BLUEPRINT OF A SYNTHETIC ORGANISM:

The Fusion of Structure, Hybrid Energy, and Decentralized Intelligence

by Sulien Valentino Solovyov

Table of Contents

Table of Contents (cont.)

Appendix:

A Word From the Author

Come, gentle reader, and let us
sit awhile beside the quiet hearth
of curiosity. What follows is not
the work of a scholar trained
in robotics or materials, nor a
repository of secret formulas, but
the musings of one who has long
watched the world and wondered
how its creatures endure. How,
I ask, might a machine bear its
burdens as life does, gathering
strength from its surroundings and
bending without breaking?

The ideas here—the fascia network, the weighted nodes, the octopus principle—are little lanterns cast upon that question. They are drawn from the rhythms of nature, from the clever economy of life, and from the simple thought that perhaps human devices might learn to move with a similar grace. These pages make no claim to invention; they are offered, quietly, for those who would wander this path of inquiry, test these notions, or simply reflect alongside them.

If a single thought here sparks
further exploration, if it prompts
a question or inspires a small
experiment, then these humble
pages have served their purpose.
May they guide the mind gently,
like a winding path through a
familiar wood, toward places
where engineering and life meet,
and where wonder is welcome.

A Note on Formatting and Intention

Readers engaging with this volume will immediately observe a significant presence of white space and wide margins. This is neither a printing error nor a stylistic flourish designed for pagination alone. It is, instead, a deliberate and foundational design choice tied directly to the book's conceptual purpose.

This text, on Bio-Inspired Hybrid Energy Harvesting Systems, is not a final scientific report or a completed technical manual.

It is a conceptual blueprint—a speculative framework for an entirely new architectural and computational paradigm.

In the spirit of the author's approach—that of the generalist mind seeking to open a new lane of inquiry—we have formatted this book as a working manuscript.

The generous margins are intentional, dedicated space provided for the reader:

1. **For Annotation:** To record immediate thoughts, critiques, and cross-references.
2. **For Sketching:** To draft alternative designs, material specifications, and circuit diagrams.

3. For Development: To allow
   the process of reading to
   become the process of active,
   creative development.

The low word count per page
ensures that each concept is
isolated, giving the reader the
necessary intellectual space to
immediately process and respond
to the ideas presented without
distraction.

We invite you to engage with these
pages as a canvas. Your insights in
the margins are, in our view, the
true next iteration of this work.

Foreword: The Failure of the Joint

Every complex system we build—
from the smallest robotic actuator
to the longest bridge span—
shares a fundamental, fatal flaw:
fragmentation.

We treat structure, power, and
intelligence as separate concerns,
layering them onto a monolithic
design. We build **joints**—ball
joints, hinges, axles—which
enable motion but simultaneously
introduce points of concentrated
weakness, wear, and eventual
failure. We bolt on **batteries**,
treating energy as a finite,
consumable resource rather than

a continuous, ambient input. We route complex **wiring**, creating centralized bottlenecks for data and control.

This traditional approach—a century of magnificent, yet profoundly segmented, engineering—is reaching its limit. Our robots are constrained by battery life, our infrastructure demands constant, costly maintenance, and our sensors perish when their isolated power sources deplete. The crisis of the modern machine is not one of material strength or computing speed; it is a crisis of **integration**.

This book offers a blueprint for a profound technological shift—a complete rejection of the fragmented machine.

The key to this revolution lies not in silicon or in larger batteries, but in a deeper understanding of **life**.

Nature builds systems where structure, power, and intelligence are seamlessly fused. It knows no "joints," only fluid, durable **fascia** that distributes stress and enables resilience. It supports no central command, instead employing **a distributed nervous system**—like the octopus—where local intelligence enables immediate, adaptive response.

And, it harvests energy not with a single mechanism, but continuously, from every minute deformation.

What you hold in your hands is the manual for building the next generation of integrated systems. We demonstrate the practical science behind the **Fascia Principle**—a hybrid network that performs **Structural Dual-Functionality**, replacing the traditional joint while simultaneously absorbing impact and generating power. We unveil the physics of **Amplified Hybrid Harvesting**, leveraging weighted, oppositely charged nodes to convert pervasive, low-level stimuli—vibrations, friction, even urban noise—into usable energy with an efficiency boost of 200-300%.

Crucially, we detail the integration of **Microcomputing at the Node,** transforming a mere power mesh into a living, decentralized nervous system capable of real-time control, fault tolerance, and autonomous management.

This is more than a technical design; it is an argument for a new paradigm. By embracing the integrated, self-sustaining logic of the biological world, we move beyond the mechanical failures of the past. We stop building machines that die when their battery runs down, and start building systems that are **alive, aware, and inherently self-sustaining.**

The era of the traditional joint is over. The era of the intelligent, energetic skin has begun.

Part 1

Chapter 1: True Bio-Inspired Design: The Resilient Network

Our engineering models, for all their complexity, remain primitive when compared to the simplest living organism. We are masters of segmentation: we design a steel frame for **structure,** a separate battery pack for **power**, and a brittle array of wires for **data**. This patchwork approach is not only inherently inefficient; it is actively fighting the laws of physics and biology.

We rely on **fragmented energy recovery**—systems that demand specific, large-scale motions (like a footfall or a major joint flexion) to generate a useful charge. They ignore the constant, pervasive, low-level stimuli that saturate every environment: the micro-vibrations from a distant highway, the subtle sway of the wind, the low-frequency hum of urban noise. This is energy left on the table, and it is the reason our autonomous devices remain enslaved to the charging port.

Biological systems, however, are models of continuous, holistic flow. The **lymphatic** system manages energy and fluid distribution not with pumps and isolated tanks, but with a decentralized, pressure-driven network. Most critically, the body's support is provided by **fascia**—the continuous sheet of flexible, resilient connective tissue that surrounds muscles, organs, and bones. Fascia is the ultimate integrated material: it provides **structural integrity**, distributes kinetic and vibratory energy, and facilitates movement without resorting to a single, high-wear failure point like the traditional joint.

Bio-Mimicry as Core Advantage

Our goal is not merely to copy nature's appearance, but to leverage its fundamental, time-tested principles to solve persistent engineering failures. This is the core tenet of **True Bio-Inspired Design**: we aim to create systems with **inherent efficiency, resilience, and adaptability** that far surpass conventional materials.

Conventional engineering prioritizes a simple, often brittle, strength-to-weight ratio. Biological resilience, however, is found in the ability to **distribute stress** and **adapt to continuous input**. A spider web's strength lies not in the breaking force of a single strand, but in its ability to manage and redirect energy across the entire geometric network.

By designing from this biological imperative, we replace the centralized fragility of mechanical systems with the decentralized durability of a living organism. Our solutions become intrinsically more energy efficient because they are always listening, always harvesting, and always adapting to their environment.

The Integrated Solution: Structural Dual-Functionality

The singular innovation of this work is the concept of **Structural Dual-Functionality**. We reject the notion that a material must be exclusively structural or exclusively power-generating.

Our **fascia-like network** is a multi-functional composite mesh that performs two primary roles simultaneously:

1. **Load-Bearing and Impact Durability:** It provides the necessary tensile strength, flexibility, and shock absorption at points of high stress and motion.

2. **Continuous Energy Harvesting:** It actively converts mechanical and electrostatic forces into usable electricity across its entire surface area, through the hybrid synergy of piezoelectric, triboelectric, and static charge mechanisms.

This integration is a game-changer for autonomous systems. The very act of the system moving, deforming, enduring impact, or even simply existing in a noisy environment, becomes a continuous source of power. The network is no longer a passive substrate for the machine; it is the **active, energetic skin** of the system itself.

The Fundamental Shift: Replacing the Mechanical Joint

The ultimate consequence of Structural Dual-Functionality is the fundamental shift it mandates: **the retirement of the traditional mechanical joint.**

For centuries, joints —the ball joints, the hinges, the axles—have been necessary evils. They enable motion but introduce the central points of wear, heat generation, maintenance, and structural weakness.

The bio-inspired, fascia-like network replaces these components with a continuous, flexible energy matrix. In a robotic arm, this network can provide the same degrees of freedom as the shoulder, elbow, and wrist, while also distributing force, absorbing impact, and simultaneously harvesting power from the motion. The system gains agility and durability, while shedding the complexity and inherent wear of traditional mechanical architecture.

This marks the definitive transition from machines built of separate, discrete components to **integrated, self-sustaining networks**. It is the realization of true bio-mimetic engineering, where structure, energy, and, as we shall see, intelligence are finally inseparable.

Chapter 2: The Failure of Centralization: From Ball Joints to the Octopus Principle

In Chapter 1, we established that conventional engineering fails by fragmenting structure and energy. But the flaw runs deeper still, extending into the system's most critical elements: its points of articulation and its method of control. If a system cannot move efficiently and think locally, it will always be brittle, inefficient, and slow.

At least that was my starting point, which is now several generations behind the rigid framed robots that seem to be doing very well for themselves. But, I still think there is room for this technological evolutionary path. So, I request that the reader stick with me. Maybe there is room for multiple engineering success stories. Or maybe, the thought experiment itself will point to a third and better path. Let me show you where I started:

The Problem of the Traditional Joint

For all their utility, traditional mechanical joints—from the simplest hinge to the most complex ball joint or axle—are engineering compromises.

They serve their purpose of enabling motion, but they introduce a catastrophic set of vulnerabilities:

1. Wear and Stress Concentration: Every traditional joint is a high-wear component. Motion concentrates stress and friction into a single, small area, requiring lubrication, generating heat, and necessitating eventual replacement. The failure of the joint is often the failure of the entire machine.

2. Complexity and Weight:

To manage these stress points, conventional designs require additional components: specialized bearings, reinforcement plates, support structures, and the external motors/actuators needed to drive the joint. This radically increases system **complexity, weight, and assembly cost**.

3. Fragmented Functionality:

The joint is structurally *separate* from the energy source and the sensor network. It is a passive,

dumb component that only
consumes power, offering nothing
back to the system's power budget
or intelligence.

In essence, I saw the mechanical
joint as a legacy weak point that
breaks the chain of structural
integrity and continuous energy
flow.

The Necessity of Pervasive, Low-Level Harvesting

The failure of the joint is
compounded by the failure of our
power architecture. Our current
power sources are designed to
capture large, infrequent jolts
of energy—a deep flexion of a

knee or a major load shift. They are oblivious to the ocean of low-level, pervasive, and ambient stimuli that surround every machine:

1. Micro-vibrations: The subtle hum of machinery, the tremor of traffic, or the ripple of wind across a surface.

2. Acoustic Energy: Sound itself, particularly low-frequency environmental noise, is a form of mechanical energy waiting to be converted.

3. Environmental Friction: The slight rubbing of air, dust, or clothing against a surface.

To achieve true autonomy, a system must escape **the tyranny of the battery and become its own perpetual power source**.

This requires a shift from harvesting only **peak events** to continuously **harvesting ambient noise**. The fascia-like network, with its hybrid combination of piezoelectric (capturing vibrations), triboelectric (capturing friction), and static charge agitation (capturing electrostatic fields), is engineered precisely to tap into this pervasive, low-level energy and convert it into a **continuous electrical baseline.**

The Computational Leap: The Octopus Principle

Even a perfectly powered, structurally integrated system would fail if its intelligence remained centralized. When a robot or piece of smart infrastructure relies on a single central processing unit (CPU) to gather all sensor data, compute a response, and then issue commands, the result is inherent latency, wiring complexity, and catastrophic **single-point-of-failure**.

Nature solved this problem millions of years ago, perhaps most elegantly in creatures like the **octopus**. Its nervous system is decentralized, with two-thirds of its neurons located in its arms. Each arm can sense, process, and act with a high degree of autonomy, even when severed from the brain. The central brain manages goals, but the local nodes handle real-time execution.

This is the inspiration for the **Computational Leap** in the fascia network: introducing **distributed microcomputing directly into the nodes**.

The nodes, already serving as structural junctions and energy conversion points, are equipped with miniaturized microcomputers. This creates a true **decentralized nervous system** within the synthetic "skin" of the machine.

This architecture enables:

1. Local Data Processing:
Each node processes data from its immediate vicinity—strain, temperature, vibration frequency—in real-time, without querying a distant CPU.

2. Adaptive Response: A node
can instantly adjust its own energy harvesting parameters (e.g., tweaking the triboelectric contact force) or modify its structural stiffness based on localized stress,

enhancing both power generation and resilience.

3. Fault Tolerance: If one node fails, its neighboring nodes pick up the slack, preventing total system collapse and ensuring continuous functionality.

By fusing structure, power, and decentralized intelligence into a single interconnected network, we move beyond the mechanical failures of the traditional joint and embrace the resilient, adaptive autonomy of the biological world. The network is not just a self-powered material; it is **a self-aware, distributed processor.**

Part 2

Chapter 3: The Fascia Network: Structural Dual-Functionality and Joint Replacement

Part 1 established the necessity of moving beyond fragmented design by adopting the bio-mimetic principles of structural and energetic integration. Chapter 3 begins the technical deep dive, detailing the core material science that makes this fusion possible: the **Fascia-Like Piezoelectric Network** itself.

This is the **Hybrid Engine**—a multi-functional composite that enables both motion and power generation through a material, rather than mechanical, solution.

The Multi-Directional Mesh Composition

The core of the invention is a **multi-directional mesh** that mimics the interconnected, non-uniform matrix of biological connective tissue. Its structural resilience comes from its anisotropy and its choice of flexible, durable materials:

1. The Structural Elements:

The mesh strands (ranging from 0.1–0.5 mm thick) are composed of flexible polymers such as **silicone, polyurethane, or carbon nanotube composites**. These materials provide the high deformability needed for flexible articulation and impact absorption.

2. The Energy Components:
Embedded directly within these flexible strands are **the piezoelectric materials** (e.g., PVDF, PZT, or electroactive polymers). Crucially, these piezoelectric elements are strategically placed in a **non-uniform pattern** across the strands and nodes. This non-uniformity maximizes strain-induced energy capture from varied, low-level mechanical inputs like flexion, torsion, and micro-vibration, ensuring the network is always "listening" to ambient stimuli.

3. The Matrix: The mesh forms a multi-directional matrix (5–10 strands/cm²) that allows forces to be distributed and redirected across a wide area. Like a spider web, the interconnected geometry prevents stress from concentrating at a single point, dramatically increasing **overall impact durability** (e.g., withstanding 1-2m drops or cyclic stress up to 100 MPa).

The Joint-Replacement Architecture

The strategic placement and inherent flexibility of the fascia network enable its most transformative function: **Joint**

Replacement Architecture.
Conventional joints require separate, rigid structures and motors to facilitate movement. The fascia network provides both the **flexibility and the necessary structural support** in a single component.

1. Flexible Articulation:
The intrinsic flexibility of the polymer-piezo mesh provides the necessary degrees of freedom for motion. The material itself is the articulation point, simplifying the entire design.

2. Stress Distribution: When
force is applied (e.g., a robot lifting an object), the network

distributes that stress across the entire surface, mitigating the destructive concentration of force that inevitably leads to wear in a ball joint.

3. Continuous Power Generation:
Critically, the very act of flexion, torsion, or load-bearing—the movements that would wear down a traditional joint—is converted into electrical energy. The system is structurally dual-functional: it performs the structural work and harvests power from that work simultaneously.

This architecture simplifies complexity, reduces weight, and transforms points of weakness into highly efficient, self-powering structural components.

Fabrication Methods and Strategic Placement

Achieving the micro-resolution and material integration required for the fascia network necessitates advanced manufacturing techniques and careful placement:

1. Fabrication: The polymer mesh uses **stereolithography** (3D printing) for high-resolution elastomer structures. The conductive/piezoelectric composites are often integrated via **electrospinning**, which allows

for ultra-fine fiber diameters
($0.1–1$ μm) necessary for highly
sensitive energy conversion.
Ultrasonic bonding and chemical
etching ensure strong adhesion
between dissimilar materials like
PZT and the elastomers.

2. Strategic Placement: The
network is not applied uniformly
across the entire system. It is
strategically embedded at
key points that experience high
tension, motion, or impact.

In Robotics? The network
replaces traditional mechanical
joints in the shoulder, elbow, and
wrist, providing continuous

motion while generating the power for the next movement.

In Infrastructure? It is embedded in load-bearing surfaces or bridge sections, continuously harvesting energy from low-frequency wind-induced sway or thermal expansion while providing superior stress management.

By transforming material science into a structural and energetic solution, the fascia network offers a robust foundation for the self-sustaining, intelligent systems of the future.

Chapter 4: Hybrid Harvesting: Piezoelectric, Triboelectric, and Static Charge Systems

The structural dual-functionality of the fascia network (Chapter 3) provides the ideal canvas for continuous energy recovery. This chapter details the **Hybrid Engine**—the synergistic combination of three distinct conversion mechanisms that allows the network to capture energy from virtually any ambient mechanical or electrostatic stimulus. The core advantage lies in redundancy: where one conversion method might fail (e.g., lack of friction), another succeeds (e.g., sound-induced vibration).

Piezoelectric Generation: Capturing Strain and Sound

The foundational layer of the hybrid system is piezoelectric generation, which directly converts mechanical stress into electrical energy.

1. Maximizing Strain Capture:

The piezoelectric elements (PVDF, PZT, etc.) embedded within the flexible polymer mesh are activated by any deformation. The energy harvested is maximized from the high-tension movements inherent to the network's function as a joint replacement—**flexion, torsion, and load shifts**—but it is also highly sensitive to minute inputs.

2. Acoustic Harvesting:

Crucially, this layer is tuned to capture **micro-vibrations** and **sound-induced oscillations** (e.g., 20–1000 Hz from environmental noise). This turns the otherwise wasteful energy of urban sound or machinery hum into a continuous electrical output, ensuring the network is always harvesting, even when the system is structurally static.

Triboelectric Harvesting: Material Contact and Modes of Operation

The outer shell of the network utilizes **triboelectric energy harvesting (TENG)**, which converts the energy of friction and contact into electricity via the **triboelectric effect** (charge transfer between materials upon contact).

1. Materials Selection: The network is partially coated with triboelectric materials (e.g., PDMS, nylon, aluminum) selected from opposing ends of the triboelectric series to maximize the **charge differential** upon contact, leading to higher voltage output.

2. Modes of Operation: This layer harvests energy through three primary modes:

- **Ambient Contact:** Charges generated from environmental friction (e.g., wind, dust, contact with external surfaces).

- **Self-Contact:** Designed rubbing between adjacent strands or components of the network itself during motion.

- **Environmental Stimulation:** Sound-induced micro-movements amplify TENG output by inducing rapid, tiny contacts between the charged surfaces.

Material Science of Charge: Sustaining Static Fields

The third layer introduces **static charge-generating materials** to induce and sustain localized electrostatic fields. While TENG focuses on kinetic friction, this system focuses on the controlled manipulation of static fields for energy transfer and material processing (Chapter 6).

Accumulation and Field Sustenance: Materials are selected for their high capacity to accumulate and retain charge: Amber and Glass are natural materials known for their ability to generate and hold charge via friction.

PTFE (Polytetrafluoroethylene) is a synthetic polymer valued for its high charge retention and low-friction properties in precision environments.

Synergy for Agitation: These sustained electrostatic fields are then utilized in combination with **low-frequency sonic vibrations (1–5 Hz)**. This synergy causes charged particles or internal charged elements to move or rearrange based on charge polarity, a mechanism essential for applications like precise material mixing and flow dynamics.

Structural Alternatives: The Deformable Solid Tube

A key internal component that works with these hybrid systems is the **movable charged element housed within the fascia threads**.

The Problem of the Gel: My early designs used a charged gel, which had numerous problems. The primary alternative path led to a Deformable Solid Tube solution.

Advantages of the Solid Tube: This tube is composed of a **charged microfoam material** (e.g., elastomeric foams, nanocomposites). It offers superior advantages:

1. Enhanced Sealing: The solid form prevents leakage, simplifying manufacturing and maintenance.

2. Improved Structural Integrity: The tube contributes actively to the thread's support.

3. Predictable Efficacy: Its solid, yet deformable, nature provides more predictable mechanical behavior than a viscous gel, ensuring stable, reliable energy generation through contact, triboelectric interaction, and electrostatic induction against the nodes.

Chapter 5: Amplification and Control: Weighted Nodes and Charge Dynamics

The hybrid nature of the fascia network ensures continuous energy capture, but continuous output alone does not guarantee utility. To power embedded microcomputers and actuators, the system must achieve high energy density. This chapter details the physical innovations—**the Weighted Node**—that provide a 2–3 fold increase in energy generation, along with the crucial mechanisms for managing the resultant electrical intensity.

The Innovation of Mass and Charge Differential

The node—the point where multiple fascia strands intersect—is transformed from a simple junction into a highly efficient, localized energy amplifier.

This innovation is based on two principles that maximize the energy transfer at the moment of contact:

1. Increased Mass (Weighted):
Each node incorporates a small, dense core (e.g., a metallic microsphere) to significantly increase its mass by a factor of

2–5 compared to the lighter surrounding strands.

2. Opposite Electrical Charge:

The nodes are manufactured or coated to possess an electrical charge (e.g., positive polarity) that is **opposite** to the charge of the movable element (the charged microfoam tube) within the fascia threads (e.g., negative polarity). This creates an intense, localized interaction that converts kinetic energy into electrical energy far more effectively than a uniform, neutrally weighted network.

The Physics of Amplification: Leveraging Momentum and Attraction

The combination of mass and opposite charge delivers the dramatic efficiency boost, providing the **2–3 fold increase** in energy output:

1. Momentum Amplification:

When the fascia strand is subjected to vibration, flexion, or impact, the heavier **weighted node** possesses greater inertia. This difference in mass relative to the lighter strand amplifies the magnitude of the force and momentum at the junction point. This amplification—rooted in the principle that force scales directly with mass and acceleration— drives greater deformation of the piezoelectric material and increases the friction at the triboelectric interfaces.

2. Electrostatic Attraction: The opposite charges on the node and the movable charged element create a **strong attractive force**. As the charged element moves past or contacts the node, this electrostatic attraction amplifies the local stress exerted on the **piezoelectric elements**. This powerful attraction (which scales with the magnitude of the charges) also increases the contact force between triboelectric surfaces, maximizing the charge transfer during the interaction. The node, therefore, acts as **a localized kinetic and electrostatic intensifier**, translating subtle, low-level movements into substantial, usable electrical power.

Controlling Charge Dynamics: Managing the High-Energy Environment

Operating an intensely charged, hybrid harvesting system necessitates sophisticated management of the electrical environment, specifically to handle high voltages and prevent damage to internal microelectronics.

1. Electrostatic Discharge (ESD) Management: The continuous, high-efficiency generation of charge by the triboelectric and static layers creates a risk of sudden, destructive **Electrostatic Discharge**. The network incorporates specialized mechanisms and materials to safely channel and dissipate excess charge. This ensures that the high-voltage environment generated for optimal energy harvesting does not create unintended disruptions or cause component failure.

2. Grounding and Shielding Systems: To maintain operational safety and protect the embedded microcomputers, the system integrates advanced grounding and shielding technologies. These are crucial for two reasons: **a)** they prevent external electrical interference from impacting the network's function, and **b)** they ensure that the internal high-voltage fields are contained, preventing accumulation to dangerous levels.

3. Optimizing Flow in Controlled Environments:

For high-precision industrial applications (e.g., material processing, Chapter 7), controlling charge is not just about safety, but about **efficacy**. By modulating the system's overall charge and leveraging its sonic-vibration capabilities, the network can precisely manipulate the flow of charged particles (e.g., in coating or sorting applications), ensuring optimal performance while maintaining a stable energy balance.

The successful implementation of the weighted node and its corresponding charge management framework elevates the fascia network from a proof-of-concept to a truly viable, high-power, self-sustaining system.

Part 3

Chapter 6: The Decentralized Nervous System: Microcomputing at the Node

The integration of the fascia network—a system that is structurally sound and energetically autonomous—sets the stage for the third, most advanced pillar of this invention: **the integration of intelligence**.

A self-powered system must also be a self-aware system. To achieve this, we must break away from the last vestiges of centralization and embrace a decentralized computational architecture.

The Octopus Principle: Architecture of Distributed Control

Traditional robotic control relies on a central processing unit (CPU) that operates in a bottleneck pattern: **Sense -> Transfer Data -> Compute ->Actuate**. This process is inefficient, slow, and fragile.

The solution is the Octopus Principle, which takes inspiration from the decentralized nervous system of creatures like the octopus. The octopus can control the complex, fluid movements of its arms with minimal input from its central brain because most of the neural processing occurs locally within the limb itself.

In the fascia network, this principle is realized by embedding a **microcomputer** directly into every **node**. This creates a truly **distributed computing architecture** where intelligence, energy generation, and structural sensing are inseparable.

Architecture Breakdown:

Local Data Processing: Each microcomputer is equipped to process sensor data relevant only to its immediate structural vicinity. It monitors strain, temperature, vibration frequency, and local charge generation. By handling these thousands of micro-calculations locally, the system bypasses the need to transfer vast amounts of raw data to a central controller, drastically reducing latency and energy consumption.

Node-to-Node Communication: Nodes communicate locally with their nearest neighbors, allowing for rapid, coordinated response and data sharing. This forms a mesh network that facilitates the flow of both energy and information.

Fault Tolerance: The failure of any single node does not cripple the system. The neighboring nodes can automatically detect the failure and adapt to route energy and control around the damaged section, ensuring continued structural support and function. This resilience is impossible in a conventionally wired, centralized system.

Adaptive Response Strategies: Real-Time Self-Optimization

The integration of local processing enables a level of **adaptive response** previously unattainable. The fascia network can instantaneously and autonomously optimize its own performance based on real-time environmental conditions.

The microcomputer at the node is designed to control and fine-tune its local energy harvesting parameters:

1. Harvesting Parameter Adjustment: If a node senses high-frequency acoustic energy (e.g., machinery noise), the microcomputer can instantaneously adjust the contact force or pressure on the local triboelectric interface to maximize charge transfer, leveraging the low-level stimulus more effectively.

2. Structural State Feedback:

The node constantly processes strain gauge data. If high, continuous stress is detected, the microcomputer can communicate to adjacent actuators to slightly shift the load, or more simply, it can prioritize the electrical energy generated from that stress point to immediately power a local function, converting destructive stress into productive power.

3. Environmental Sensitivity: In dynamic environments (e.g., wind in an infrastructure application), the nodes can coordinate to adjust the network's effective stiffness or elasticity momentarily. This not only optimizes the piezoelectric strain capture but enhances the overall stability and longevity of the material itself.

Energy Management: Distributed Power Balancing

In traditional systems, a central power management unit decides when and where to allocate finite battery power. In the fascia network, energy management is a distributed, cooperative process managed by the network of nodes.

1. Monitoring and Optimization: Each node monitors the energy harvested locally and optimizes its own power consumption. If a node is currently under high strain (and thus generating high output), it communicates this status to its neighbors.

2. Distributed Power Balancing: This allows for instantaneous, democratic distribution. Energy is channeled precisely where it is needed.

- If a remote node is performing a heavy actuation task and requires more power, the entire network—not just a central battery—participates in sourcing that power.

- Energy storage modules embedded throughout the network are managed locally by the nodes, minimizing transmission loss and ensuring that power is always available at the point of action.

Optimizing Node-Level Consumption:
The microcomputers are programmed to operate in an ultra-low-power state, "waking up" only when significant structural or environmental input is detected.

Their own power requirements are minimal and are typically met by the energy they harvest immediately, maintaining a cycle of self-sufficiency.

By combining the **Fascia Principle** with the **Octopus Principle**, the system achieves a state of self-governance, transforming the machine from a segmented collection of parts into **a self-aware, resilient, and energetically autonomous entity**.

Part 4

Chapter 7: Sonic and Electrodynamic Synergy in Manufacturing

The fascia network's ability to generate intense, controlled electrostatic fields, combined with its capacity to detect and generate specific vibrations, extends its utility far beyond structural support and power generation.

This chapter details how this synergy becomes a **precision tool** for advanced industrial processing and material manipulation, solving complex challenges in high-value manufacturing.

Vibrational Agitation:
Amplifying Electrostatic Forces

The key to precision is the controlled use of **low-frequency sonic vibrations** (specifically in the 1–5 Hz range) to amplify and direct electrostatic forces.

1. Amplification Principle: While electrostatic fields (sustained by materials like PTFE and Glass) create the potential for movement, the mechanical vibrations provide the kinetic energy necessary for continuous, controlled agitation. This **low-frequency oscillation** enhances **micro-friction**, causing particles to constantly overcome their static adhesion and move along the electric field lines.

2. Controlling the Flow: The frequency of the vibration, paired with the material's size and charge polarity, dictates the flow pattern. This allows the system to achieve highly predictable particle displacement and organization, ensuring homogeneous blending or precise segregation.

This synergy allows for **material manipulation without mechanical intervention** (e.g., stirring paddles or pumps), drastically reducing wear, contamination, and energy consumption in the manufacturing process.

Flow Dynamics and Industrial Efficacy

The marriage of sonic and electrostatic control yields immediate, high-value applications in particle and fluid management:

1. Particle Sorting:
Electrostatic forces, coupled with agitation, enable the precise separation of particles based on subtle differences in their charge and size—a capability crucial in industries like **mining, electronics, and pharmaceuticals**, where purity is paramount.

2. Spray Coating and Deposition: By charging fluid

droplets and using electrostatic fields, the network can ensure uniform coverage of surfaces. The forces promote the movement of the charged droplets toward a target, improving deposition efficiency and reducing material waste in processes like **additive manufacturing and high-precision coating**.

3. Electrostatic Precipitation:

In gas stream filtration (e.g., removing fine particulate matter), electrostatic forces attract charged particles to collection plates. The controlled fields enhance the efficiency of this process, supporting environmental and air quality applications.

High-Precision Applications: Nanoscale Manipulation

The ability to control material interaction at the micro-level translates directly into breakthroughs at the nanoscale, driving innovation in materials science:

1.Nanoparticle Manipulation and Alignment: Electrostatic fields and sonic vibrations offer the precision required to control the movement and alignment of nanoparticles and molecules. This capability is vital for creating new materials with specific structural arrangements, such as enhancing the strength and durability of composites by ensuring uniform distribution of carbon fibers.

2. Blending in Pharmaceuticals:
In the production of powders and coatings, ensuring uniform distribution of active ingredients is non-negotiable. Electrostatic manipulation achieves this blending consistency without the shearing forces that can damage delicate compounds.

3. Fluid Agitation and Dispersion: In liquid mediums, the synergy facilitates the dispersion of materials and prevents clumping, critical for creating uniform composite materials and stable liquid pharmaceutical suspensions.

By offering a self-aware, energy-efficient method of material manipulation, the fascia network fundamentally enhances processing precision, pushing the frontiers of Nanotechnology and Materials Science.

Chapter 8: Market Revolution: Applications of the Self-Sustaining System

The fascia network is not merely an incremental improvement; it is a platform technology that fundamentally alters the cost, durability, and computational models across high-value markets.

By providing structure, energy, and intelligence in a single, resilient form, the self-sustaining system creates opportunities for unprecedented product autonomy and efficiency.

Robotics: The Self-Sufficient Body

The transformation of robotics is immediate and profound. By integrating the fascia network, robots move past the constraints of limited battery life and prone-to-failure mechanical joints.

1. Enhanced Autonomy:

The continuous, ambient energy harvesting ensures that operational lifecycles are dramatically extended, making truly autonomous missions—in deep-sea exploration, space surveying, or remote infrastructure maintenance—practical for the first time.

2. Flexibility and Durability:

The Joint-Replacement Architecture allows for seamless, fluid motion while distributing stress and absorbing impact, making robots inherently more resilient to damage. This is essential for manufacturing lines (reduced downtime) and dynamic environments (increased survivability).

3. Decentralized Locomotion:

The microcomputers at the nodes enable distributed control, leading to advanced, rapid, and adaptive control strategies that mimic the agility of biological organisms.

Wearable Devices & Medical Implants: Fusion with the Human Body

The fusion of energy generation, flexibility, and biocompatibility creates a new paradigm for devices attached to, or placed inside, the human body.

1.Battery-Free Wearables:
The network, embedded in smart clothing and exoskeletons, harvests energy directly from the wearer's movement, friction, and heat. This eliminates the need for bulky, heavy batteries, drastically improving comfort, aesthetic design, and user compliance.

2. Life-Extending Medical Devices: The system's biocompatible nature is perfect for implantable devices (e.g., pace-makers, continuous glucose monitors, sensors). By harvesting energy from the body's minute movements and fluid dynamics (mimicking the lymphatic system), these devices can become self-powering, eliminating the need for invasive battery replacements and extending functional longevity.

3. Personalized Processing: Localized nodes enable personalized data processing, allowing a smart garment to instantly analyze muscle strain or movement metrics without lagging

communication to a smartphone or centralized unit.

Infrastructure: Self-Powered Smart Skin

For large, static structures, the fascia network acts as an intelligent, self-powered "skin" that generates its own electricity while providing advanced structural health monitoring.

1. Sustainable Power Source:
Integrated into the structural components of bridges, buildings, and wind farms, the network continuously harvests energy from low-frequency vibrations, wind sway, and thermal expansion.

This sustainable power source can directly feed sensors, LED lighting, and communication arrays.

2. Advanced Structural Health Monitoring (SHM): The distributed microcomputers at the nodes provide continuous, high-resolution strain and stress data. By processing this locally, the network can detect and report subtle damage or fatigue instantly, increasing the resilience and longevity of critical infrastructure while reducing costly, manual inspection cycles.

Automotive & Aerospace: Efficiency and Resilience

The need for light weight, high durability, and extreme efficiency in the transport sectors makes this technology indispensable.

1. Improved Fuel Efficiency (Automotive): Integrating the network into vehicle suspensions and seats harvests energy from vehicle motion and road vibration, reducing reliance on the alternator and improving overall fuel economy. The material's energy-absorbing capabilities enhance safety and dampening performance.

2. Weight Reduction (Aerospace): Replacing heavy, centralized wiring and battery systems with a continuous, lightweight, self-powering structural network is paramount. The system can be integrated into aircraft wings and fuselage to harvest energy from aerodynamic forces while reducing overall structural mass, improving fuel efficiency, and enhancing maneuverability.

3. Inherent Resilience: The distributed, fault-tolerant nature of the fascia network ensures that if one area is compromised (e.g., impact damage to a vehicle component), the rest of the

network continues to function
without a cascading system failure.

Conclusion: The Inevitable Integration

The work presented in this book is the result of a fundamental challenge to the industrial age's most entrenched assumptions. For decades, our technological progress has been defined by fragmentation: discrete systems for structure, power, and control.

This model—represented by the failure-prone mechanical joint and the finite battery—has reached its philosophical and practical limits. The **Bio-Inspired Hybrid Energy Harvesting System** offers the inevitable alternative: **Integration**.

The Merging of Biology and Engineering

This system moves beyond superficial bio-mimicry to leverage the core principles of biological survival:

1. Structural Dual-Functionality: Replacing inert structural materials and failure-prone joints with active, resilient, energy-generating **fascia**. The material itself performs the work of both support and power.

2. Amplified Hybrid Efficacy: Achieving self-sustaining power through **redundant harvesting** (piezoelectric, triboelectric, static

charge) amplified by the deliberate physics of the **weighted node**. This ensures continuous energy capture from the ambient noise of the environment.

3. Decentralized Intelligence:
Adopting the **Octopus Principle** by fusing microcomputers directly into the structural **nodes**. This creates a truly autonomous, self-aware, and fault-tolerant network that manages its own power and processes information locally.

Unparalleled IP Breadth and Future Autonomy

The claims made within this text—spanning material

composition, energy conversion, computational architecture, and manufacturing process—constitute a formidable competitive barrier and define a new technological platform. The **system's unparalleled IP breadth** ensures its dominance across multiple high-value markets.

The future of technology is the future of autonomy. Robots will no longer require umbilical cords or docking stations. Infrastructure will self-power its sensors and self-report its fatigue. Medical devices will extend their functional life indefinitely within the human body.

We have demonstrated that the highest form of engineering is not the mastery of complexity, but the return to simplicity—to the elegant, integrated design perfected by evolution. The result is a structure that is not merely controlled, but **aware**; not merely powered, but **energetic**; and not merely built, but **alive**.

The era of separate components is finished. The era of the **self-sustaining, intelligently controlled structure** is here.

Author's Afterword: The Starfleet Mindset

An Enduring Curiosity: From Blueprint to Starfleet

Come, gentle reader, you have walked with me through a wilderness of theory, tracing the outlines of a technology that, as yet, exists more fully in the mind than in the metal. The pages you have just turned are not a final technical manual but a conceptual blueprint, offered in the same spirit that science fiction offers tomorrow's reality today. The work here—the fascia networks, the weighted nodes, the octopus principle—are little lanterns cast upon the question of how a machine might endure its burdens as life does.

For inspiration, we need look no further than the great speculative visions of our time, such as those that gave us Star Trek.

The original communicator, a miraculous, pocket-sized device for instantaneous connection, was once pure fantasy; now, it is a cultural artifact rendered obsolete by the very mobile device in your hand.

Similarly, the android Data, with a synthetic body that was structurally sound, self-repairing, and integrated with decentralized intelligence, embodies the ultimate form of the integrated, self-sustaining system we have explored. The difference between the fragmented machine of today and a Data-level synthetic organism is precisely the shift from centralized fragility to decentralized durability. That chasm is what this book seeks to bridge.

I am a generalist , not a specialist, and my contribution is less about solving a known problem and more about opening a new lane of inquiry. The central purpose of publishing this speculative framework is to map the areas of technological synergy that, to my knowledge, remain critically unexplored in current engineering literature.

I submit that the following 10 areas, when viewed as a cohesive, integrated system, represent the true, unanswered challenges and the most fruitful paths for future exploration:

1. Structural Dual-Functionality (Fascia Principle): The use of a single composite material to simultaneously perform load-bearing/impact durability and continuous energy harvesting, thereby retiring the traditional mechanical joint.

2. Continuous, Pervasive Low-Level Harvesting:
The architectural design to systematically capture energy from continuous, ambient, low-level stimuli (micro-vibrations, sound, wind sway) that conventional systems ignore.

3. Hybrid Conversion Redundancy:
The synergistic combination of piezoelectric, triboelectric, and static charge mechanisms within a single mesh to ensure energy capture redundancy across varied environmental stimuli.

4. Amplification via Weighted Nodes:
The use of small, dense,

internally weighted nodes to physically amplify the force and momentum transferred from low-level movements to the energy conversion materials.

5. Electrostatic Attraction Amplification: The manufacturing principle of giving nodes and internal elements opposite electrical charges to create a strong, localized attractive force, which amplifies mechanical stress on piezoelectric elements and increases triboelectric contact force.

6. Microcomputing at the Node (Octopus Principle): Embedding a fully capable microcomputer directly into every structural/ energetic node to create a genuinely decentralized, mesh-networked nervous system.

7. Real-Time, Localized Self-Optimization: The ability of an individual node's microcomputer to instantaneously adjust its own energy harvesting parameters (e.g., tweaking triboelectric contact force) based on local, real-time environmental input (e.g., acoustic energy).

8. Adaptive Structural State Feedback: The node's capacity to process local strain data and convert destructive stress into productive power by prioritizing

the generated electrical energy for
an immediate local function.

9. Distributed Power Balancing:
An autonomous, democratic
system where every node monitors
its local output and coordinates
with neighbors to channel
power precisely where needed,
eliminating the need for a central
power management unit.

**10. Sonic and Electrodynamic
Synergy in Manufacturing:**
The controlled use of low-
frequency sonic vibrations
(1–5 Hz) to amplify and direct
sustained electrostatic fields for
highly precise particle sorting and

material manipulation without mechanical intervention.

These are the starting points—the rough navigational charts—for building systems that are alive, aware, and inherently self-sustaining. I invite you to take up this canvas. May your insights in the margins be the true next iteration of this work.

Technical Analysis and Implementation Roadmap: Bio-Inspired Hybrid Energy Harvesting System (BHEHS)

This blueprint details the technical implementation phases, current technology readiness, and fundamental differentiation of the Bio-Inspired Hybrid Energy Harvesting System (BHEHS). The BHEHS fundamentally re-architects autonomy by merging structure, power, and decentralized intelligence.

1. Required Technical Implementation Steps

The realization of the BHEHS requires a phased, multidisciplinary approach focusing on advanced material science, microelectronics, and control architecture integration.

Phase 1: Materials Science Optimization (TRL 5-7)

A. Technical Focus: Composite Fabrication & Efficacy

Deliverable: Develop robust, flexible, high-yield composite materials (e.g., PVDF-elastomer/ CNT nanocomposites) via electrospinning and additive manufacturing.

Challenge: Ensuring structural integrity while maximizing the strain-induced charge derived from the material.

B. Technical Focus: Tribo/Static Charge Elements

Deliverable: Optimize triboelectric material pairings from opposing ends of the series for maximal charge differential. Stabilize charge retention in static materials (PTFE, glass).

Challenge: Perfect the manufacturing of the charged

microfoam tube for predictable
internal movement and energy
transfer.

**Phase 2: Microelectronics
Integration (TRL 4-6)**

**A. Technical Focus: Node
Microcomputer Design**

Deliverable: Develop ultra-
low-power, high-efficiency
microcomputers capable of
local data processing (strain,
temperature, vibration frequency)
and energy management.

Challenge: These must fit within
the extremely constrained volume
of the structural node while
maintaining performance.

B. Technical Focus: Communication Protocol

Deliverable: Design the high-speed, local mesh network protocol for decentralized node-to-node communication.

Challenge: Ensuring real-time adaptive response and robust fault tolerance across thousands of networked nodes.

Phase 3: System Integration and Control (TRL 3-5)

A. Technical Focus:
Amplification Validation

Deliverable: Rigorously test and validate the 2–3 fold energy amplification claim derived from the weighted, oppositely charged node design.

Challenge: Validation must occur under complex, cyclic stress and vibration profiles that mimic real-world use.

B. Technical Focus: Adaptive Control Algorithms

Deliverable: Develop firmware that enables nodes to autonomously adjust local harvesting parameters (e.g., triboelectric contact force) based on real-time environmental input, realizing the Octopus Principle.

Challenge: Integrating the feedback loops between energy generation and computational management in a seamless, low-latency manner.

C. Technical Focus: Manufacturing Scaling

Deliverable: Establish high-throughput, integrated manufacturing processes (e.g., high-resolution 3D printing of elastomers with simultaneous electrospinning of active elements).

Challenge: Successfully scaling the fascia network structure while maintaining micron-level precision and functional integrity.

2. Proximity to Implementation and Timeline

The BHEHS is a near-term, high-potential technology.
While the full integrated system is speculative, component technologies are mature:

A. Existing Technology Readiness (TRL 7+):
Piezoelectric and triboelectric nanogenerators (PENG/TENG) are routinely demonstrated in laboratory environments. Flexible

polymer composites and ultra-low-power microcontrollers are commercially available.

B. The Integration Gap (TRL 3-4): The primary challenge is not invention but integration and scale. Merging these diverse technologies into a durable, cohesive, structurally load-bearing material (eliminating the joint) is the key hurdle.

Estimated Timeline and Deliverables

A. Laboratory Demonstration (1–3 Years): Proof-of-concept prototypes validating the structural dual-functionality and the 2–3x amplification claim on small-scale components (e.g., a robotic finger or simple structural beam).

B. System-Level Integration (3–5 Years):
Development of the full decentralized microcomputer network and the creation of adaptive control algorithms for full-scale systems like a robotic arm or infrastructure monitoring segments.

C. Industrial Deployment (5–10 Years):
Robust, high-reliability commercial products in low-risk sectors (e.g., smart textiles, non-critical infrastructure monitoring) before moving into high-load

applications (robotics, aerospace).

3. Differentiation from Current Robotics Practice

The BHEHS represents a philosophical and architectural break from current state-of-the-art engineering.

A. Architectural Feature: Structural Integrity & Motion

i. Current Robotics Practice:
Fragmented design relies on rigid materials connected by high-wear mechanical joints (hinges, axles).

ii. BHEHS Differentiation:
Integrated & Resilient design employs Fascia-Like Networks

for Structural Dual-Functionality, eliminating the joint and distributing stress across the entire system.

B. Architectural Feature: Energy Source & Management

i. Current Robotics Practice:
Centralized & Finite power relies primarily on a single, heavy battery for power, requiring periodic downtime for recharging.

ii. BHEHS Differentiation:
Decentralized & Continuous power harvests Ambient Stimuli (sound, micro-vibration, friction) across the entire body, using the system's own movement as a continuous power source.

C. Architectural Feature: Computational Model & Intelligence

i. Current Robotics Practice:
Centralized CPU funnels data from all sensors to a single main processor, creating latency and a single point of failure.

ii. BHEHS Differentiation:
Distributed Processing employs the Octopus Principle with microcomputers in every node, enabling Local Data Processing, real-time adaptive response, and robust fault tolerance.

D. Architectural Feature: Material Function

i. Current Robotics Practice:
Passive materials are inert; they only provide support and consume power.

ii. BHEHS Differentiation:
Active & Self-Aware materials act as a sensor, a generator, a structural component, and a processing unit simultaneously.

www.ingramcontent.com/pod-product-compliance
Lightning Source LLC
Chambersburg PA
CBHW040927210326
41597CB00030B/5208